Climate Change

In the News™

Corona Brezina

16
EasyRead Large

RHYW

Copyright Page from the Original Book

Published in 2008 by The Rosen Publishing Group, Inc.
29 East 21st Street, New York, NY 10010

First Edition

Library of Congress Cataloging-in-Publication Data

Brezina, Corona.
Climate change / Corona Brezina.—1st ed.
p. cm.—(In the news)
Includes bibliographical references and index.
ISBN-13: 978-1-4042-1913-7 (lib. bdg.)
ISBN-10: 1-4042-1913-7 (lib. bdg.)
1. Climatic changes. 2. Global warming. 3. Greenhouse effect—Atmospheric. I. Title.
QC981.8.C5B726 2008
363.738'74—dc22

2007006129

Manufactured in the United States of America

On the cover: *(Clockwise from top right)* Floodwaters flow past the Opera House in Dresden, Germany; a polar bear crosses sea ice pools; the smokestacks of a paper mill release carbon dioxide into the atmosphere.

Set in 16 pt. Verdana

TABLE OF CONTENTS

Introduction

Image 1

Global climate change is one of the most crucial issues that humanity will have to address during the twenty-first century. The potential consequences of climate change extend far beyond higher temperatures caused by global warming. Climate change could devastate wildlife and transform the oceans and land. It could disrupt agriculture and increase the intensity of weather events.

Just recently, people have grown aware of how human activity could bring about warming trends. In the 1970s and 1980s, some scientists still believed that the earth could be on the brink of entering a new ice age. After it was shown that average temperatures were actually rising steadily, it took some time before scientists could prove that humans were responsible for the increase. Skeptics still protested that fears of global warming and climate change were alarmist. They claimed that any fluctuations were part of the earth's natural cycles, not a consequence of human activity.

A deluge of scientific research on climate change in the last two decades has quieted most of the skeptics, however. Studies repeatedly have confirmed that numerous changes occurring in the atmosphere, biosphere, oceans, and other natural systems are unprecedented. Scientific evidence supports the scenario in which human activity—in particular, the emission of carbon dioxide (CO_2) and other gases—causes the atmosphere to warm up, which in turn influences the global climate in a variety of ways.

World leaders have not come up with a viable plan to counter climate change. In order to curb the emission of CO_2 and other gases that cause global warming, humans would have to cut back on their use of fossil fuels such as coal and oil. That's not easy to do in a world with more than six billion inhabitants, many of whom enjoy the comforts and convenience provided by electricity, automobiles, and manufactured goods. Ideally, scientists and leaders will succeed in devising a solution that will benefit both humanity and the natural world.

1

The Warming Earth

Image 2

The earth's climate is changing, and the evidence is readily apparent. Since the 1990s, record-setting high temperatures have been recorded in places across the world. In the Arctic, polar bears are struggling for survival as their icy habitat shrinks. At the opposite pole of the earth, a chunk of the Larsen B ice shelf of Antarctica that was larger in area than Rhode Island collapsed into the ocean in 2002. Twenty-seven thousand people died in Europe of heat-related causes during the brutal summer heat wave of 2003. The exceptionally intense 2005 hurricane season—which spawned Hurricane Katrina and devastated New Orleans and the Gulf Coast—raised awareness of the possibility that global warming could contribute to extreme weather events. Behind these recent developments is a slight but persistent and relentless rise in average global temperatures.

Taken alone, any of these events or phenomena might be dismissed as a fluke or explained away as an extreme but isolated instance of the earth's natural cycles. Viewed as part of a larger picture and considered in the context of long-term trends and patterns, however, the evidence of global warming is indisputable. What's more, there is a solid and growing majority consensus among the scientific community that it is human activity driving the temperature increase.

A Comfortable Climate

Climate and weather are not the same thing. Weather is a day-to-day phenomenon. Climate is the average weather over a long period of time. Meteorology—the study of the weather—is an imprecise science. Forecasts are uncertain: they predict the likelihood of precipitation or a possible range of high and low temperatures, and they grow less accurate as they look further into the future.

Climate science—or climatology—is a far broader and more complex area of study. Climatology encompasses longer time periods than meteorology. It draws upon climate records from the past as well as data about present-day conditions. Scientists examine changes that are undetectable in our daily lives but may provide indications of significant long-term shifts in the climate. While the average person may take the cyclical turning of the seasons for granted, scientists study climate patterns as the product of interactions among numerous forces.

Image 3: An ice crack cuts through Antarctica's Larsen ice shelf. In 2002, a huge area of the ice shelf collapsed into fragments.

Even small changes, though, can have an impact on the organisms that live within a given climate. Plants, animals, and microorganisms often take their cues from the temperature. They depend on regular levels of moisture for survival. If their ecosystem is disrupted by climate change, it may force them to adapt to new conditions in order to survive.

Humans are also dependent on a stable climate. Drastic changes in regional climates, such as desertification or persistent flooding, could force massive evacuations from affected areas. Climate change could affect agriculture and contribute to shortages of fresh water. Like other animals experiencing a change in climate, human beings may have to alter their behavior and settlement patterns in order to survive.

Tracking Climate Change

Image 4: Researchers examine an ice core extracted from Siple Dome in Antarctica. This core provides climatic information extending back 80,000 to 100,000 years.

The earth's average temperature is about 59 degrees Fahrenheit (15 degrees Celsius). The climatic conditions of the present day have remained fairly constant for about 8,000 years, a span of time longer than the recorded history of the human race. Considering this long period of stability, it's easy to see why many people find it difficult to believe that the climate is headed toward drastic change, if not in our lifetimes, than during those of the next generation. This conception of a reliable climate, though, is an illusion. Through the eons, the earth's climate has experienced a roller-coaster ride of shifts that scientists are still struggling to piece together and explain.

Unfortunately for researchers, evidence about past climates becomes sparser the further they reach back in time. The oldest data is provided by analysis of seabed cores collected from the bottom of the ocean. These samples yield information about past temperatures and ocean conditions going back as far as sixty-five million years. Another source of valuable data comes from ice cores drilled in Antarctica and Greenland, which provide scientists with time capsules of past climates. The Antarctic ice core, extracted at a research base called Vostok Station, is 11,775 feet (3,589 meters) long and took eight years to drill. As ice accumulated through the ages, each layer trapped dust and bubbles of air within the crystals. Within that trapped material, information can be obtained about that era's climate. The ice core has so far revealed climatic information going back 400,000 years, though the entire record may stretch back 720,000 years or even longer.

Ice and seabed cores give a general sense of how the earth's climate has changed over eons. Other types of evidence reveal more specific information about conditions during more recent eras or specific geographic regions. The age of organic matter can be determined by a process called carbon-14 dating, and some mineral formations such as stalagmites can be dated through radiometric dating. Preserved trees—as well as long-lived live trees such as the bristlecone pine—can provide a climate record for a region going back thousands of years. Paleobotanists examine tree ring growth, which tells the story of how growing conditions change throughout time. Similarly, samples of ancient pollen, which are highly resistant to decay, give clues about the diverse plant life that thrived during certain periods.

The geological record shows that the earth has been cycling through ice ages for the past 2.5 million years. Vast ice sheets advanced to cover much of Europe and North America, then receded during shorter warm periods called interglacials. The last ice age began the transition into an interglacial 18,000 years ago. The earth's climate is still in the midst of that interglacial.

The periodic recurrences of ice ages demonstrate climate change on a geological timescale hard for humans to fathom. In 1920, a Serbian mathematician named Milutin Milankovich announced that he had found a pattern that fit the timing of the glacial periods. He hypothesized that the onset of ice ages had been initiated by variations in the earth's positions in space. According to his theory, three overlapping cycles affect the amount of solar radiation reaching the earth. The first involves the varying elliptical shape of the earth's orbit around the sun. The second concerns the extent of the earth's tilt on its axis. The third involves the "wobble" of the earth's axis. These cycles last about 100,000 years, 42,000 years, and

22,000 years, respectively. When the Milankovich cycles, as they are called, coincide to minimize the amount of solar radiation reaching the earth, conditions are favorable for ice ages.

The amount of radiation emitted by the sun itself is inconstant as well. This variation generally is attributed to sunspot activity, which changes over the course of an eleven-year cycle. Catastrophic events also may have triggered drastic climate change in the earth's distant past. Examples include massive volcanic eruptions, which spew carbon dioxide into the atmosphere, and asteroid collisions such as the one often credited with causing the extinction of the dinosaurs sixty-five million years ago.

Global Warming

In the context of these extreme climate changes that have taken place throughout the earth's history, it's understandable that some people question the causes of present-day global warming. Isn't it possible that climate change today is part of the earth's natural climatic variation?

In 1988, the United Nations formed the Intergovernmental Panel on Climate Change (IPCC), made up of about 2,500 scientists from across the world, to investigate the issue. In the group's First Assessment Report, released in 1990, researchers reported that the earth's average surface temperature had risen about 1°F (0.56°C) over the previous century, although they could not prove that the change was a

Image 5: Every time a volcano erupts, it spews carbon dioxide and other gases into the atmosphere.

direct result of human activity. By the time the IPCC released their Third Assessment Report in 2001, they had determined that temperature increases of the twentieth century were the greatest of the entire millennia, and they cited strong evidence that human activity was largely responsible for the change. The report also projected that temperatures would rise an additional 2.5° to 10.4°F (1.4° to 5.8°C) by the year 2100.

Scientists based the report on information gathered from many sources: the geological record, recent climate data collected from across the world, the results of field research, and climate models used to predict future trends. These computer climate models mathematically simulate the interactions of the atmosphere, oceans, land surface, biosphere, and other factors that influence climate. A credible climate model can project how a change in one factor, such as increasing atmospheric carbon dioxide (CO_2), would impact future conditions. Climate models have limitations, since climatic processes are incredibly complex. Scientists don't claim to understand every aspect of how the climate works. Models are constantly being refined in order to take more factors into account.

If the projections are accurate and global warming is well underway, some people may wonder, is that really so bad? After all, it wouldn't seem that a slight increase in temperature should affect humans too adversely. They may think that it could even be a positive thing, resulting in longer summers and growing seasons and less harsh winters. The truth is that the consequences of global warming are likely to extend far beyond rising temperatures. Climate change brought about by global warming could radically transform some aspects of the world we now take for granted.

2

The Greenhouse Effect

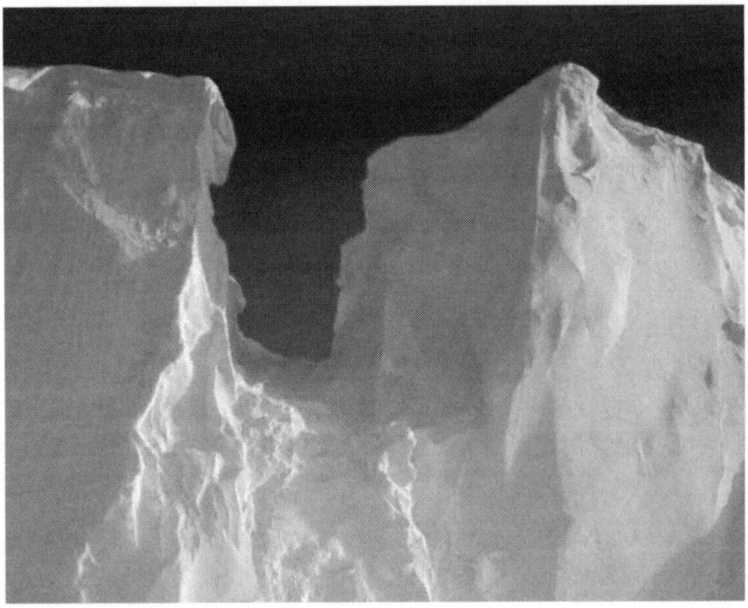

Image 6

Life as we know it could not survive on the earth without the phenomenon called the greenhouse effect. Energy from the sun reaches the earth in the form of visible light. The atmosphere allows this shortwave radiation to pass through to the earth's surface, which warms when it absorbs some of the energy. The earth reradiates this energy outward in the form of infrared radiation.

Certain components of the atmosphere—greenhouse gases—block some of this long-wave radiation from streaming back into space, however. It is trapped in the earth's atmosphere, where it keeps the lower atmosphere relatively warm. Without the greenhouse effect, the earth would be frozen in a never-ending ice age.

Global warming is caused by increased levels of greenhouse gases. Greater concentrations of these gases trap a greater proportion of the infrared radiation given off by the earth. As a result, the temperature of the atmosphere rises, thereby impacting the global climate. Put basically, temperatures are rising steadily because more and more greenhouse gases are accumulating in the atmosphere and trapping heat.

Carbon Dioxide (CO_2)

The greenhouse gas most associated with global warming is carbon dioxide (CO_2). When scientists talk about rising levels of carbon in the atmosphere or carbon emissions, what they're referring to is carbon existing in the form of CO_2.

Carbon dioxide is crucial to life on the earth because of its role in the carbon cycle, which circulates carbon in and out of the atmosphere. During respiration, animals take in oxygen (O_2) and expel CO_2. Plants, in turn, take in CO_2, incorporate the carbon into their structure, and release oxygen, which is in turn breathed in by animals, and so on. In this way, plants and animals contribute to the maintenance of a balance of gases in the atmosphere.

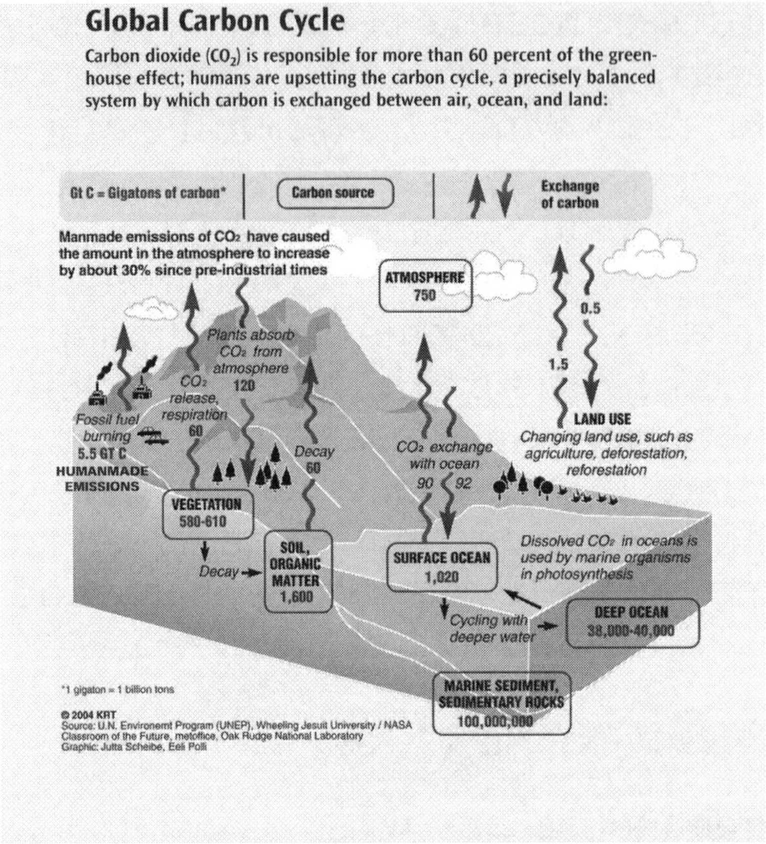

Image 7: This graphic illustrates the cycle of carbon moving through the atmosphere, hydrosphere (oceans), and geosphere (land). Billions of tons of carbon move through these three "reservoirs" every year. Five-and-a-half billion tons of that total are created by human activity.

Respiration and photosynthesis are the key processes of the carbon cycle, but there are other components at work as well. Carbon sources constantly release carbon into the atmosphere, while carbon sinks—such as oceans and forests—constantly absorb it and remove it from the atmosphere. CO_2 is continually dissolving into the ocean and leaving the ocean in about equal amounts. Limestone and other carbon-containing rocks release CO_2

as they are weathered by the elements. At the same time, some of the carbon in the ocean sinks to the bottom and begins the process of forming rock, thereby removing itself from cycle activity. These are only a few of the carbon sources and sinks involved in the carbon cycle.

In the course of the carbon cycle, carbon moves through three reservoirs: the atmospheric reservoir, the ocean reservoir, and the land reservoir (which includes carbon contained in the soil as well as in living organisms). Proportionately, though, very little of the world's total carbon is actively involved in the carbon cycle. Most of it is locked into rocks and fossil fuels such as oil, coal, and natural gas. This is called the geological reservoir of carbon.

Through mining and drilling, humans are drawing some of this vast geological reservoir of carbon up from the earth. Combustion (burning) of fossil fuels converts stored carbon into atmospheric carbon. In this way, human activity is influencing the natural balance of the carbon cycle by adding significant amounts of CO_2 to the atmosphere that otherwise would have remained trapped in the earth's surface.

Human Contributions of Carbon Dioxide

It might seem that knowledge of a link between global warming and carbon emissions is a recent development, but it is not a new idea to scientists.

Image 8: Chemist Charles Keeling conducted pioneering research on the earth's climate by monitoring carbon dioxide (CO_2) levels at the Mauna Loa observatory in Hawaii.

After the dawn of the Industrial Revolution, humans began relying more and more heavily on fossil fuels for energy. In 1895, the Swedish chemist Svante Arrhenius completed the first research suggesting that increased atmospheric carbon dioxide could cause global temperatures to rise. He believed that this might be a positive development, benefiting agriculture and making the world a more comfortable place for humans.

Arrhenius's ideas remained untested and unproven for many decades, since nobody yet monitored carbon dioxide levels in the atmosphere. In 1958, a young chemist named Charles Keeling proposed that the U.S. Weather Bureau measure atmospheric carbon dioxide levels at a new observatory located on the Hawaiian volcanic mountain of Mauna Loa. The result of his measurements over time is the famous graph known as the Keeling curve. The resultant line is shaped like a sawtooth because the overall amount of carbon dioxide in the air is lowest in the summer and rises during the winter. Overall, despite seasonal dips, the graph shows a trend of steadily increasing CO_2 levels through the years.

In 1979, the National Academy of Sciences set up a panel called the Ad Hoc Study Group on Carbon Dioxide and Climate to evaluate the possibility that global warming was occurring due to human emissions. The members concluded that rising CO_2 levels would indeed result in climate change. The panel also warned of Earth's built-in time delay: it could be decades before changes already underway would result in measurable evidence of warming. This meant that the world didn't have the luxury of waiting and seeing. If it hoped to prevent significant and potentially catastrophic climate change it needed to act now before all the evidence could be gathered and theories confirmed by observable facts. Once evidence of global warming became apparent, it would be too late to do anything about it. Yet most world governments took no action to address the problem when the alarm was first raised.

Since that time, data from ice cores has confirmed the link between rising levels of greenhouse gases and increased temperatures. Over the past 400,000 years, warm periods have coincided with the highest levels of CO_2 and methane, another greenhouse gas.

So, to what extent has human activity raised atmospheric CO_2 levels? Carbon dioxide makes up such a small proportion of the atmosphere that amounts are measured in the parts per million (ppm). The preindustrial level of CO_2 measured about 280 ppm. By the time Keeling began his monitoring, it had risen to 316 ppm. As of 2005, the figure measured 378ppm. Archaeological evidence reveals that during no other time in the past 650,000 years—the range of the Antarctic ice core record and long before humans first appeared on the earth—have CO_2 levels surpassed 300 ppm. Climate models indicate that atmospheric CO_2 will only continue to rise dramatically due to human activity, so human beings are heading into uncharted climatic territory. Before the end of the twenty-first century, they will probably be living in a world that is warmer than any humans have ever experienced before.

Other Greenhouse Gases

Carbon dioxide is not the only greenhouse gas causing global warming. A handful of other pollutants also contribute to the greenhouse effect. These gases are present in the atmosphere in much lower concentrations than CO_2, but they are more effective at trapping heat. The measurement of this property is called the global warming potential (GWP). CO_2 has a GWP of 1. Methane has a GWP of 21, meaning that a certain amount of methane has 21 times the effect as the same amount of carbon. The GWP of nitrous oxide is 310.

Even water vapor acts as a greenhouse gas. In an effect called water-vapor feedback, warmer air retains more moisture, which in turn increases water's contribution to the greenhouse effect. This would imply that as global temperatures rise and more ocean and lake water evaporates, greater amounts of water vapor in the air would accelerate warming. On the other hand, water vapor forms clouds. Clouds can block solar radiation, which would reduce warming, but they can also trap heat in the lower atmosphere, raising earth's surface temperatures. The overall role of water vapor in global warming is one of the uncertainties of climate science.

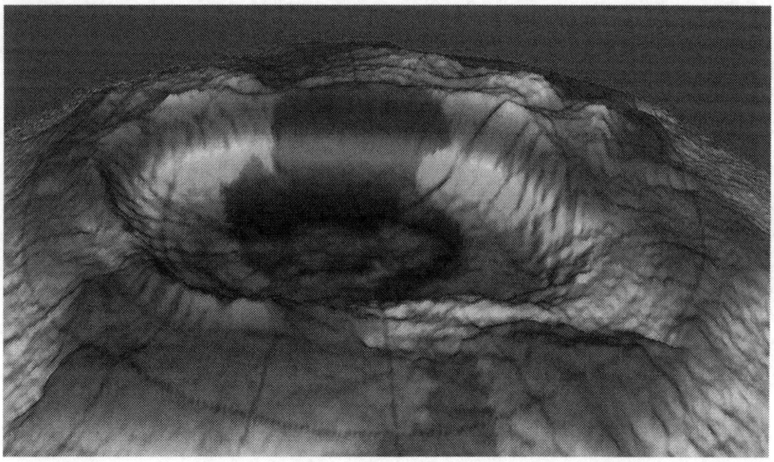

Image 9: Industrial use of CFCs, a potent class of greenhouse gases, was phased out after they were found to be creating a hole in the protective ozone layer surrounding the earth. This is a 3-D NASA image of the ozone hole.

Methane is the primary component of natural gas, which is used as a fuel. Methane is produced through bacterial decomposition. Methane-releasing microbes live in environments such as swamps and rice paddies. They also reside in the digestive systems of termites and animals such as cows. Over a third of the methane added to the atmosphere through human activity is emitted by livestock. Although it is a potent greenhouse gas, it stays in the atmosphere for a shorter period of time than CO_2.

Nitrous oxide, sometimes called laughing gas, is released by bacteria. The volume of nitrous oxide emitted as a result of human activity now surpasses that of natural sources. Nitrous oxide is a by-product of industrial production, of burning fossil fuels and biomass, and of agricultural processes.

Methane and nitrous oxide both occur naturally in the atmosphere, though human activity has increased their concentration. Several of the most potent greenhouse gases, however, are created solely through human manufacture. These include perfluorocarbons (PFCs), chlorofluorcarbons (CFCs), hydrofluorocarbons (HFCs), and sulfur hexafluoride. Sulfur hexafluoride, the most potent known greenhouse gas, has a GWP of 23,900.

In the 1980s, it was discovered that CFCs were causing depletion of the ozone layer, the component of the atmosphere that screens the earth from the sun's harmful ultraviolet rays. Governments across the world worked together to pass regulations banning the use of CFCs, used mainly in refrigeration and air-conditioning. Since then, global emissions of CFCs have declined dramatically. This success story provides hope that the world's leaders will be able to work together in the future to curb other greenhouse gases.

3

Consequences of Global Warming

Image 10:

Sometime in the not-so-distant future, Montana's Glacier National Park may have to be renamed. Established in 1910, the park once boasted about 150 glaciers. Climate change has reduced that number to fewer than 30, and they're melting quickly. Scientists say that the park's glaciers will be gone within a few decades.

Melting Down

The retreat of Montana's glaciers is a phenomenon that is being duplicated around the world. As glaciers recede to higher altitudes, where the air is cooler, mountain ranges such as the Himalayas and the Andes are seeing their ice coverings diminish. Scientists predict that most of the glaciers of the Alps will vanish by 2050. In Iceland, volunteer observers have recorded the steady receding of their glaciers since the mid-1990s. At the southern tip of South America, ice fields in some parts of traditionally harsh and forbidding Patagonia have given way to dry land dotted by lakes.

Meltdown is occurring in the permafrost, the layer of permanently frozen ground below the surface of Alaska, Siberia, and other far northern regions of the world. The top layer of the permafrost, called the active layer, melts enough during the summer to allow plant growth. Dead plants sink below the active layer before fully decomposing. Deeper in the lower layers of the permafrost, such partially rotted organic matter is suspended in a perpetual deep freeze.

Scientists have measured recent temperature increases of 3° to 6°F (1.7° to 3.3°C) in the Alaskan permafrost. Symptoms of the warming can be seen in collapsing buildings and "drunken trees" that tilt over as the hard permafrost underneath turns mushy. The main concern of most climate scientists, though, lies beneath the ground. If the deeper layers of the permafrost begin to melt, the collected organic matter frozen for millennia could thaw and begin to decompose. This process could release huge amounts of methane and CO_2, greenhouse gases that would further perpetuate global warming.

Colder climates such as alpine or subarctic zones are especially sensitive to the effects of warming. More than any other place in the world, the earth's poles are critically vulnerable to global warming. If the average temperature were to increase just 5°F (2.8°C), the equator would only grow slightly warmer, but temperatures would rise by about 12°F (6.7°C) at the poles.

Image 11: In Alaska, the melting of permafrost beneath this tennis court has caused it to buckle.

At the North Pole, the Arctic Sea is covered by an ice cap with an average thickness of less than 10 feet (3 m). The sea ice is thinning, though, and its total area is shrinking. In 1979, 1.7 billion acres (6.9 million square kilometers) were covered by ice year round. Since then, the total coverage has declined by about 250 million acres (1 million sq km).

Diminishing sea ice brings additional climatic consequences due to a change in albedo, the reflectivity of a surface. Sea ice, which reflects much of the solar radiation that falls on it, has a high albedo. Open water, on the other hand, has a very low albedo: it absorbs far more heat from the sun than ice does. When radiation from the sun warms the open water, it increases the melting of nearby ice, which in turn decreases the coverage of high albedo ice and increases the coverage of low albedo water, resulting in more radiation absorption and ice melt. This vicious cycle is called the ice-albedo positive feedback.

Since the Arctic ice cap floats on open water, its melting does not raise sea levels. At the South Pole, however, Antarctica poses a different scenario. The Antarctic ice cap is 10,000 feet (3,048m) thick, and it contains about 70 percent of the world's freshwater. The Antarctic continent has been pressed down beneath sea level by the great weight of the ice sheet. Therefore, its melting would contribute to a potentially catastrophic rise in sea levels.

In early 2002, scientists were flabbergasted to observe huge chunks breaking off the Larsen B ice shelf on the Antarctic peninsula. Within thirty-five days, the entire ice shelf—an area of 1,268 square miles (3,284 square kilometers)—had collapsed. The incident caused no far-reaching environmental impact, but it served as a warning sign of unsuspected instability in the region. Since then, scientists have begun tracking other disturbing changes in the Antarctic ice.

Antarctica is not the only ice-covered landmass of particular concern to climatologists. Greenland, which contains 8 percent of the world's freshwater in its ice sheet, could potentially have an enormous environmental impact if meltdown accelerated. The signs are troubling. Scientists report that Greenland's temperatures have risen, its ice cap has receded, and the frequency of "glacial earthquakes" caused by lurching ice has increased.

Oceans Under Threat

If the ice sheet covering Greenland were to melt, sea levels across the world would rise 23 feet (7m). The East Antarctic Ice Sheet (considered more vulnerable than the larger West Antarctic Ice Sheet) contains enough water to raise sea levels about 28 feet (8.53m). If all of the earth's ice coverings thawed, sea levels would increase by 230 feet (70m). Total meltdown, however, would take millennia.

There are factors besides melting ice that affect sea levels. The most significant is the effect of thermal expansion. Warmer water takes up more space than cooler water. If ocean temperatures increase due to global warming, sea levels will rise due to the expansion of the water. The IPCC estimated in its 2001 assessment report that by 2100, sea levels will have risen between 4.3 and 33 inches (10.9 and

83.8 centimeters). Most of this projected figure is attributed to thermal expansion, not melting.

Image 12: Flowing melt water cascades off of Greenland's ice sheet. In recent years, the summer melt season has come earlier and been more extensive.

Melting in Greenland could have consequences beyond its contribution to rising sea levels. The currents of the oceans loop around all seven continents to form a huge circuit called the Great Conveyor. This system keeps the global climate stable by circulating warm water from the equator to the poles. The North Atlantic portion of the Great Conveyor—known as the Gulf Stream—is critical to regulating the climate of Europe. London, for example, is farther north in latitude than Lake Superior. Due to the heat brought north by the Gulf Stream, however, Britain's climate is moderated and more temperate than lower-latitude northern Michigan's.

The Gulf Stream is propelled by a pattern called thermohaline circulation *(thermo meaning* "heat," and *haline* meaning "salt"). After crossing the Atlantic, warm, salty water loses heat through evaporation and, because it is now colder and therefore heavier, sinks down to the ocean floor between Greenland and Norway. From there, the cooler current flows southward and eventually rises back to the surface as it becomes heated closer to the equator.

A change in either ocean temperature or salinity could affect thermohaline circulation, and both factors could be impacted by global warming. It has been shown that the ocean temperatures are rising. If Greenland's ice sheet were to melt, freshwater would pour into the North Atlantic Ocean, thus decreasing the salinity. Slowing or shutting down the Gulf Stream could cause parts of western Europe to cool despite the larger trend toward global warming.

The oceans act as the world's most powerful and effective carbon sink. Atmospheric CO_2 levels would be much higher if the oceans did not take in some of the excess released by human activity. But this increased absorption of CO_2 is causing a harmful shift in the chemical composition of the oceans. Dissolved CO_2 raises the levels of carbonic acid, which decreases the pH level of the ocean. This acidification of the oceans could have serious consequences for marine life.

Unexpected Weather

Climate records from past centuries and millennia indicate that sudden shifts in the global climate can contribute to extreme weather events and cause dramatic—sometimes even catastrophic—changes in regional climates. The hypothetical cooling down of Europe due to the breakdown of the Gulf Stream would be one such result. Climate science is so complex, however, that there is uncertainty and disagreement about the likelihood of various possible scenarios.

It is likely that global warming will cause shifts in rainfall patterns. Throughout the twentieth century, average precipitation increased worldwide. Although climate models differ on whether this trend will continue, instances of extremely heavy rainfall are expected to increase, bringing with them more frequent and severe flooding. At the same time, higher temperatures may lead to greater evaporation of soil moisture in some regions, thus exacerbating drought and triggering increased wildfire activity.

Drought and other factors are likely to affect the availability of freshwater across many parts of the world. In mountainous regions, snow pack provides an important source of freshwater. Water supplies from snow pack will diminish as temperatures rise. Freshwater will be more subject to contamination due to tainted floodwater or increased salinity resulting from rising sea levels.

Some climate models predict that global warming will influence the behavior of weather events such as hurricanes, possibly increasing their frequency and/or intensity. Warmer surface waters of the ocean provide more energy to fuel hurricanes. Even a slight increase in temperature can contribute to a hurricane's intensity. Although there is not yet a consensus on the issue, many scientists attribute a trend of increased hurricane activity since 1995 to the effects of global warming. Global warming may impact the patterns of typhoons, cyclones, monsoons, and the El Nino-Southern Oscillation (ENSO), a cyclical weather phenomenon of the southern Pacific Ocean.

Changing Landscapes

In a hundred years, the effects of climate change might be apparent merely by comparing an early twenty-second century world map with one from the beginning of the twenty-first century. Rising ocean levels could erase the edges of low-lying islands and coastal areas. A greater area of inland territory could become vulnerable to flooding and perhaps remain under water semipermanently.

In a few cases, entire countries could be threatened by rising sea levels. Half of the Netherlands—one of Europe's so-called Low Countries—already lies at or below sea level. Higher sea levels and intensified flooding could overwhelm the country's extensive system of dikes, holding ponds, and other means of defense against encroaching water. Poorer countries such as low-lying Bangladesh in Southeast Asia, which relies heavily on agriculture, could be devastated. Bangladesh already experiences frequent severe flooding due to monsoons, and higher sea levels would bring even greater destruction. Small islands, such as the nation of Maldives in the Indian Ocean, could lose much of their dry land area.

Image 13: In low-lying countries such as the Netherlands, rising sea levels due to climate change could overwhelm dikes and other means of holding back the water.

The ocean coasts of the world would be inundated by rising sea levels, as would delta regions such as the Mississippi river delta and the Nile river delta of Egypt. Much of Florida would be swamped. New York,

New Orleans, San Francisco, Shanghai, Bangkok, Tokyo, and other coastal cities would be jeopardized by encroaching water levels.

A map of the world's climatic zones will probably look very different in 100 years. Deserts would be drier, prolonged drought could bring about desertification in some previously fertile regions, and some forested areas could give way to grasslands. Alpine climate areas would be reduced, and northern climates would be warmer. Maps representing ocean temperature and acidity would change to reflect alterations in these categories.

The geological record indicates that past global climate shifts—which were sometimes accompanied by rising sea levels—often saw the extinction of species that could not adapt to changing environmental conditions. Humans also have periodically had to respond to lesser climate pressures like regional droughts or cold spells. The next century will likely test how a world dominated by humans will respond to large-scale climate change and whether the human race will take appropriate steps to guarantee its survival. If the global community doesn't take aggressive measures to reduce concentrations of greenhouse gases, human beings may have to rely on forced migrations and adaptation to new climatic conditions if they hope to avoid extinction.

4

Living in a Warmer World

Image 14

The world's coral reefs are suffering. Corals are living entities that provide habitat for a range of marine life of such biodiversity that reefs are often called an oceanic version of a tropical rain forest. It is a fragile ecosystem that has long sustained damage from pollution, overfishing, and natural disasters.

The most recent threat to coral reefs is climate change, and it could prove to be the deadliest. A phenomenon called coral bleaching, caused by higher ocean temperatures, turns vibrant reefs into chalky skeletons. Reefs can recover from short periods of warming, but longer heat waves are often fatal. In 1998, an exceptionally hot year killed about 16 percent of the world's coral reefs. If temperatures continue to rise, the earth's reefs could be devastated. In addition to warming, climate change threatens coral reefs through acidification of ocean water. Corals use a compound called calcium carbonate as a building material for reefs. As ocean acidity rises, it causes a drop in levels of calcium carbonate.

The plight of coral reefs is emblematic of how climate change is impacting the natural world. Ecosystems most sensitive to warming are already being transformed. As temperatures continue to rise, human communities will likely begin to feel the effects of climate change as well.

Wildlife on the Brink

Image 15: In some areas of the Arctic, polar bears struggle for survival as their habitat experiences dramatic changes due to global warming.

Wildlife across the world is experiencing the pressures of habitat loss and degradation of natural ecosystems. As humans convert open, untamed land into villages, towns, cities, suburbs, farm fields, and industrial sites, plants and animals are pushed out. Every day around the world, some species disappear forever and others inch closer to extinction. Climate change, which further transforms land and weather conditions, adds to the strain on wildlife. Most at risk are species living near the poles, around high-altitude alpine regions, and in other climates currently experiencing particularly drastic increases in temperature.

In the Arctic, the effects of climate change already are apparent in the plight of the polar bears. Polar bears spend much of their year on the sea ice. They hunt seals that surface to breathe near holes in the ice, and female polar bears dig snow caves where they give birth to their young. Around Canada's Hudson Bay—the southern edge of the polar bears' range—permanent sea ice has been receding, and the seasonal ice breakup has been starting two or three weeks earlier in the summer. These changing conditions have taken a toll on the health of polar bears living in this region. Average body weight has dropped by about 15 percent, and the bear population has decreased. Instances of triplet births, once common, are now a rare occurrence. Polar bears are powerful swimmers, but they are having to swim longer distances between ice floes in their search for food. In

2005, scientists discovered polar bears drowning as they crossed open water.

On the Antarctic peninsula, Adelie penguins are experiencing the impact of climate change. Adelie colonies continue to thrive farther south in Antarctica, but on the northern fringes of their habitat, temperatures are rising and Adelie populations are drastically falling. Gentoo penguins, which live in more temperate regions, are moving south into the Adelie's range. One factor in the Adelie penguins' decline may be their dependence on bare sea ice, especially during nesting season. Adelie penguin eggs cannot be incubated in snow or slush. Greater expanses of open water leads to increased evaporation and snowfall, creating adverse conditions for Adelie eggs.

More and more species are migrating due to climate change affecting their native habitats. Some species, such as the Gentoo penguin, are expanding their range. This phenomenon also is seen on mountains, where plant growth is encroaching upward into territory that was formerly too cold to support it. More often, species are driven out of their habitats and have no place to go. For more than a decade, ecologist Camille Parmesan has been tracking the range of the Edith's checkerspot butterfly, which was once found throughout parts of northern Mexico and southern California. Due to rising temperatures, the butterfly's original habitat no longer supports the snapdragon (a flowering herb) that is the food source for Edith's checkerspot caterpillars. The butterfly is now extinct in 80 percent of its original range, and it cannot migrate up the California coast because of San Diego's urban sprawl.

The case of the Edith's checkerspot butterfly and the snapdragons demonstrates how the impact of climate change on one species can have a cascading effect on other wildlife within the same ecosystem. Climate change is disrupting the timing of natural cycles, which creates further imbalances in the interactions between species. In the Netherlands, a certain type of caterpillar has adapted to the earlier arrival of spring by hatching sooner. The flycatcher birds that feed their chicks on the caterpillars, however, have not adjusted their migratory patterns. As a result, they miss the peak caterpillar population and are more likely to experience food shortages. Some birds and other migratory species are changing their seasonal migration patterns.

Climate change exacerbates (worsens) the threat posed by invasive species, which sometimes are transported into new habitats by humans. Often, native species have no natural defenses against invasive species, or cannot compete with them for resources such as food or light. Insect pests, in particular, may be able to extend their range into northern habitats as temperatures rise and cause damage to forests and crops.

Image 16: Intense storms such as Hurricane Katrina may become more commonplace in the future, since research indicates that climate change contributes to hurricane activity.

The Effect on Humans

For the most part, human beings have removed themselves from the day-to-day ordeals of survival in the natural world. Most of us do not harvest or forage for our own food. We keep our own climates stable through heating and air-conditioning. Most weather events are inconveniences, not catastrophes. Considering our disconnect from the habitats and ecosystems surrounding us, it is easy to see why many people find it hard to believe that climate change could affect them personally.

The potential impact of climate change on humans is a contentious point of debate. The most immediate effects would be felt by people whose homes and livelihoods could be directly harmed by changes in the environment around them, such as people who live in coastal or island communities. But climate change could potentially have economic and social costs that would be felt by everyone. Skeptics claim that the possible threat to humans posed by climate change is overstated, and that technology will provide solutions to any consequences that may occur.

Across the world, hundreds of millions of people live along the ocean's coastline and in other regions that would potentially be vulnerable to rising sea levels. Cities affected by higher water levels would be forced to undertake public works projects such as seawalls, massive water-pumping stations, and other protective measures to counter flooding. Poorer countries and countries with extremely low-lying areas might have no choice but to resort to mass evacuations.

Weather events exacerbated by climate change could bring with them increasingly heavy costs, in terms of human suffering as well as money. Hurricane Katrina caused at least $60 billion in damage to New Orleans and the Gulf Coast and killed more than 1,500 people. It will take years to rebuild New Orleans, and some people believe that the city will never be the same again. It would be disastrous for the United States if destructive storms such as Katrina occurred with greater frequency. For impoverished countries, increased incidence of such weather-related events could be catastrophic.

Image 17: A pest-control truck sprays insecticide in an attempt to eradicate mosquitoes carrying West Nile virus, which first entered the United States in 1999.

Experts agree that climate change will affect agriculture, but opinions differ sharply on the overall global impact. Some growing areas would be hurt by water shortages, shifts in local climates, extreme weather, and rising sea levels. Crop pests may become more pervasive, requiring greater amounts of chemical pesticides to control them. Other aspects of global warming, however, could benefit crops and plant growth worldwide. Warmer climates in northern regions would mean an increase in tillable cropland, and higher temperatures combined with longer growing seasons could boost yields. For some plants, atmospheric CO_2 acts as a natural fertilizer by stimulating photosynthesis.

Climate change could take a toll on human health. Higher temperatures and shorter, milder winters—conditions favorable for both plants and crop pests—also benefit insects such as mosquitoes and ticks, as well as infectious diseases spread by these and other parasites. The United States could experience an increase in exotic diseases such as West Nile virus, which first entered the country in 1999. Global warming could expand the temperature zone in which the malaria-bearing mosquito thrives. Worldwide, contaminated freshwater sources could contribute to the spread of waterborne diseases.

Climate Change Firsthand

For most of us, the effects of global warming are taking place far from our daily lives. In a few places, however, climate change is already an immediate predicament.

In the Arctic, experienced hunters have fallen through the ice in places where it had been solid for most of their lifetimes. Robins are migrating northward into places where the languages of native populations have no name for the unfamiliar bird. Residents of the far north are buying air conditioners. On the small Alaskan island of Sarichef, the village of Shishmaref may soon relocate to the mainland. In been hemmed in by sea ice before winter arrived. Today, the protective ice has been reduced and the island's permafrost is softening, leaving the village vulnerable to sea surges, erosion, and sinkholes. the past, the island had always

Image 18: Grass thrives at the shoreline of Africa's Lake Chad. Thirty years ago, this area was underwater.

Far from Shishmaref, the island residents of Tuvalu in the South Pacific Ocean are concerned that climate change may someday force them to move. The nine low-lying coral atolls that make up the nation are highly vulnerable to rising sea levels. The government has already considered possible evacuation plans if their islands become uninhabitable.

In Africa, it is believed that climate change is contributing to the shrinkage of Lake Chad, once one of the largest lakes in the world. Over the past forty years, it has been reduced to about one-twentieth of its former area. A study funded by NASA indicated that a warmer, drier climate combined with increased agricultural water and irrigation use has brought about the lake's decline. The plight of Lake Chad already has affected millions of Africans in the form of crop failures, dying livestock, and depleted fish stocks.

5

Taking Action on Climate Change

Image 19

Most climate change scenarios predict a steady rise in CO_2 emissions that in turn will cause a steady rise in global temperatures. It is generally believed that the resulting climate change would occur gradually. The geological record, however, indicates that there have been times in the past that the global climate shifted abruptly, as if a switch had been flipped. Could rising CO_2 levels caused by human activity bring about a sudden climate transformation? Scientists don't rule out the possibility. They use the term "dangerous anthropogenic interference," or DAI, in discussing hypothetical situations in which human activity causes the climate to pass a tipping point, after which significant climate change is irreversible. "Anthropogenic" refers to environmental changes brought about by humans.

One such trigger would be the collapse of the Gulf Stream. Another would be the release of methane trapped beneath the permafrost, as well as the release of gases stored in sediments on the ocean floor. A third possible trigger would be the potential breakdown of the Amazon rain forest, an important carbon sink in the global climate. Experts disagree on what threshold of CO_2 in the atmosphere is likely to bring about DAI. The most often cited figure is 500 parts per million, which is about double the preindustrial level of CO_2. Some scientists believe that the threshold is even lower.

There is a consensus among scientists that action must be taken quickly to counter global warming. The effects of CO_2 emissions do not manifest themselves immediately. Even without taking increasing emission levels into account, it will take decades for the earth's climate system to respond to CO_2 currently being added to the atmosphere. If tomorrow humans suddenly ceased all activities that spewed CO_2 into the atmosphere, global warming and climate change would still be an issue because of the enormously high levels of CO_2—which is slow to break down and dissipate—already in the atmosphere. It is far more likely that CO_2

emissions will continue to rise sharply, and, if they do, humans could be setting the stage for a global disaster that will be felt by their children and grandchildren. The obvious solution is to take immediate steps to reduce human contributions of CO_2 and other greenhouse gases to the atmosphere.

Reducing Greenhouse Gas Emissions

How to go about reducing carbon emissions, however, is a controversial matter. Slightly less than 30 percent of the CO_2 released into the atmosphere each year is caused by the burning of forestland. Most CO_2 emissions come from the use of fossil fuels in industrialized countries. The United States is responsible for nearly a third of the world's greenhouse gas production. This amount is slightly higher than the contributions of all of the countries of Europe combined, which together produce about 28 percent of the world's CO_2 emissions. Overall proportions of emissions may change in the near future, as rapidly developing China is poised to become the world's largest emitter of greenhouse gases. The Chinese built about seventy-five coal-fired power plants in 2005 and are continuing to construct more.

Reduction of greenhouse gas emissions will require the efforts of national governments, industries and businesses, regional and local governments, and individuals. Many industries are unwilling to embrace change. The powerful and highly profitable energy industry, in particular, would have to shift away from a reliance on fossil fuels. The automobile industry would have to raise the average gas mileage of vehicles and expand the use of alternative fuels. Numerous industries have balked at instituting emission-reduction measures, claiming that it would be expensive and potentially ruinous in a competitive marketplace.

Image 20: If this Chevrolet Volt electric concept car ever reaches the market, drivers of the future will be able to charge their car battery in an ordinary wall socket.

Global Measures

In 1992, two years after the IPCC released its First Assessment Report, representatives from governments across the world assembled in Rio de Janeiro, Brazil, for the United Nations Conference on Environment and Development. The conference, often called the Earth Summit, resulted in the adoption of the UN Framework Convention on Climate Change. President

George H.W. Bush signed the treaty and submitted it to the Senate, where it was unanimously approved. Under the Framework Convention, the United States and other industrialized countries would aim to limit emissions to 1990 levels.

Nevertheless, most nations failed to comply, and emissions continued to rise throughout the 1990s. In 1997, delegates reconvened in Kyoto to draft a stronger resolution that mandated cuts in emissions. The United States signed the agreement—the Kyoto Protocol—but President Bill Clinton never submitted it to the Senate for approval and passage.

During the 2000 presidential campaign, George W. Bush pledged to support federal regulations limiting CO_2 emissions if he became president. Shortly after taking office, however, he announced that the United States was withdrawing from the Kyoto Protocol, and that he was reversing his position on emissions caps. Limits on CO_2emissions would now be mostly voluntary for industries, rather than mandatory. In 2005, the Kyoto Protocol went into effect, without the participation of the United States.

In the absence of any firm national policy on curbing greenhouse gas emissions, many states and communities have taken steps on their own. More than 300 mayors across the country have approved the U.S. Mayors Climate Protection Agreement, which calls for cities to comply with the conditions of the Kyoto treaty. In 2006, the U.S. Supreme Court heard a case in which a number of states, cities, and organizations brought a lawsuit against the federal government in an attempt to require the Environmental Protection Agency to regulate CO_2 and other greenhouse gases as pollutants.

Working Toward a Sustainable World

There's no easy solution to the challenge of reducing CO_2 emissions. Somehow, the United States—along with the rest of the world—would have to find ways to lessen its dependence on fossil fuels. This could only happen through a combination of strategies.

Renewable energy sources, such as wind, solar, and geothermal power, could be used to generate a greater proportion of the United States's electricity. Auto manufacturers could produce vehicles with higher fuel efficiencies or develop vehicles that incorporate alternative fuel sources. In some cases, new technologies could be used to reduce emissions caused by burning fossil fuels. In a process called carbon capture and storage, for example, CO_2 can be pumped underground or otherwise disposed of, rather than being released into the atmosphere.

Concerned citizens can analyze their own energy use and work to reduce their "carbon footprint." A variety of books and online resources are available that describe ways ordinary people can act to counter global warming. You contribute to CO_2emissions every time you flip a light switch, drive across town, or sit down to a hamburger. Your food sometimes travels thousands of miles before it reaches your plate, and meat production requires more water and fossil fuel energy than plant crops. The raising of livestock also requires the clearing of forestland—an important carbon sink—and the releasing of large quantities of methane. Try switching to energy-efficient light bulbs, bicycling or taking public transportation more often, and preparing a meatless dinner several times a week.

There are many ways you can reduce energy use around your home and in your daily life. Your effort, however small, is important: a significant reduction in CO_2 emissions will only occur if individuals, communities, industries, and governments all take steps toward reducing fossil fuel–based energy use.

Glossary

albedo The fraction of incoming solar light that is reflected by the earth's surface.

anthropogenic Caused by human activity.

atmosphere The blanket of gases that surrounds the earth's surface.

biomass Organic matter, especially plant life, that can be used as a source of energy.

carbon cycle The processes (including photosynthesis, decomposition, and respiration) through which carbon is circulated through natural carbon reservoirs: the atmospheric reservoir, the oceanic reservoir, and the land reservoir.

desertification The transformation of habitable land into a desert.

ecosystem A community of organisms along with their physical environment.

fossil fuel An energy source derived from the decayed remains of ancient life.

greenhouse gas An atmospheric gas that contributes to the greenhouse effect by absorbing radiation that would have been reflected into space.

ice core An ice sample drilled and extracted from a glacier or ice sheet.

interglacial A period of warmer climates that occurs between two ice ages.

permafrost Ground that is permanently frozen.

radiometric dating A method of determining the age of a sample by measuring the concentration of a particular radioactive element within it.

salinity A measure of the amount of salt in a solution.
solar Relating to the action of the sun's light or heat.
sunspot A relatively dark patch on the sun's surface where the temperature is cooler than in surrounding areas.
thermohaline circulation Circulation of ocean currents driven by wind, heat, and salinity.

For More Information

The Conservation Fund
National Office
1655 N. Fort Myer Drive, Suite 1300
Arlington, VA 22209-2156
(703)525-6300
E-mail: postmaster@conservationfund.org
Web site: http://www.conservationfund.org
The Conservation Fund's "Go Zero" Carbon Emission page: http://www.gocarbonzero.org

Environmental Defense
Membership and Public Information
1875 Connecticut Ave NW, Suite 600
Washington, DC 20009
(800)684-3322
E-mail: members@environmentaldefense.org
Web site: http://www.environmentaldefense.org/home.cfm

Intergovernmental Panel on Climate Change (IPCC)
c/o World Meteorological Organization
7 bis Avenue de la Paix, C.P. 2300
CH-1211 Geneva 2
Switzerland
Web site: http://www.ipcc.ch

National Oceanic and Atmospheric Administration (NOAA)

14th Street and Constitution Avenue NW, Room 6217
Washington, DC 20230
(202)482-6090
Web site: http://www.noaa.gov

The Nature Conservancy
4245 North Fairfax Drive, Suite 100
Arlington, VA 22203-1606
(703)841-5300
Web site: http://www.nature.org

Stop Global Warming
15332 Antioch Street #168
Pacific Palisades, CA 90272
Web site: http://www.stopglobalwarming.org/default.asp

Web Sites

Due to the changing nature of Internet links, Rosen Publishing has developed an online list of Web sites related to the subject of this book. This site is updated regularly. Please use this link to access the list:

http://www.rosenlinks.com/itn/clch

For Further Reading

Archer, David. *Global Warming: Understanding the Forecast.* Malden, MA: Blackwell Publishing Professional, 2006.

Brower, Michael, Ph.D., and Warren Leon, Ph.D. *The Consumer's Guide to Effective Environmental Choices: Practical Advice from the Union of Concerned Scientists.* New York, NY: Three Rivers Press, 1999.

David, Laurie. *Stop Global Warming: The Solution Is You!.* Golden, CO: Fulcrum Publishing, 2006.

Dow, Kirstin, and Thomas Downing. *The Atlas of Climate Change: Mapping the World's Greatest Challenge.* Berkeley, CA: University of California Press, 2006.

Gaughen, Shasta, ed. *Global Warming.* San Diego, CA: Greenhaven Press, 2005.

Heinberg, Richard. *Powerdown: Options and Actions for a Post-Carbon World.* Gabriola Island, BC, Canada: New Society Publishers, 2004.

Houghton, John. *Global Warming: The Complete Briefing.* New York, NY: Cambridge University Press, 2004.

Langholz, Jeffrey, Ph.D., and Kelly Turner. *You Can Prevent Global Warming (and Save Money!): 51 Easy Ways.* Kansas City, MO: Andrews McMeel Publishing, 2003.

Bibliography

Allaby, Michael. *A Change in the Weather.* New York, NY: Facts On File, Inc., 2004.

Flannery, Tim. *The Weather Makers: How Man Is Changing the Climate and What It Means for Life on Earth.* New York, NY: Atlantic Monthly Press, 2005.

Glick, Daniel. "GeoSigns." *National Geographic,* September 2004, pp. 12–33.

Gore, Al. *An Inconvenient Truth: The Planetary Emergency of Global Warming and What We Can Do About It.* Emmaus, PA: Rodale Press, 2006.

Kolbert, Elizabeth. *Field Notes from a Catastrophe: Man, Nature, and Climate Change.* New York, NY: Bloomsbury Publishing, 2006.

Long, Douglas. *Global Warming.* New York, NY: Facts On File, Inc., 2004.

Maslin, Mark. *Global Warming: A Very Short Introduction.* New York, NY: Oxford University Press, 2004.

Montaigne, Fen. "EcoSigns." *National Geographic,* September 2004, pp. 34–55.

Morell, Virginia. "TimeSigns." *National Geographic,* September 2004, pp. 56–75.

About the Author

Corona Brezina is a writer and researcher living in Chicago, Illinois. A graduate of Oberlin College and Conservatory, she has written more than a dozen books, many of which focus on issues of science, technology, climate, and the environment.

Photo Credits

Cover, image 1, image 13, image 14 Shutterstoc k.com; cover, image 17 © Getty Images; cover (bottom) © Ty Milford/Getty Images; image 2, image 5, image 10, image 20 © AFP/Getty Images; image 3 © Greenpeace/epa/Corbis; image 4 Alexander Calhoun/National Science Foundation; image 6 Michael Van Woert, NOAA NESDIS, ORA; image 7 © MCT Gr aphics/Newscom.com; image 8 © Jim Sugar/Corbis; image 9 NASA; image 11, image 18 © Aurora/Getty Images; image 12 © Roger Braithwaite/Peter Arnold, Inc.; image 15 © Pal Hermansen/Getty Images; image 19 © Ed Viggiani/Getty Images.

Designer: Tom Forget; **Photo Researcher:** Marty Levick

Black Cover Material

NEW TITLES IN THIS SERIES

CLIMATE CHANGE

DARFUR: AFRICAN GENOCIDE

DOPING: ATHLETES AND DRUGS

GREEN TECHNOLOGY:
EARTH-FRIENDLY INNOVATIONS

NUKES: THE SPREAD OF NUCLEAR WEAPONS

OIL: THE ECONOMICS OF FUEL

Books For ALL Kinds of Readers

At ReadHowYouWant we understand that one size does not fit all types of readers. Our innovative, patent pending technology allows us to design new formats to make reading easier and more enjoyable for you. This helps improve your speed of reading and your comprehension. Our EasyRead printed books have been optimized to improve word recognition, ease eye tracking by adjusting word and line spacing as well as minimizing hyphenation. Our EasyRead SuperLarge editions have been developed to make reading easier and more accessible for vision-impaired readers. We offer Braille and DAISY formats of our books and all popular E-Book formats.

Made in United States
Orlando, FL
04 November 2026

90207391R00042